I0502700

Overall Equipment Effectiveness

Key to Improving Asset Effectiveness

Brice Alvord

Limits of Liability Disclaimer of Warranty

The author and publisher of this book and the accompanying materials have used their best efforts in preparing this program. The authors and publisher make no representation or warranties with respect to (the accuracy, applicability. fitness, or completeness of the contents of this program. They disclaim any warranties expressed or implied), merchantability, or fitness for any particular purpose. The authors and publisher shall in no event beheld liable for any loss or other damages, including but not limited to special, incidental, consequential, or other damages. As always, the advice of a competent legal, tax, accounting or other professional should be sought. The author and publisher do not warrant the performance, effectiveness, or applicability of any sites listed in this book. All links are for information purposes only and are not warranted for content, accuracy or any other implied or explicit purpose.

This manual contains material protected under International and Federal Copyright Laws and Treaties. Any unauthorized reprint or use of this material is prohibited.

ALERA Publishing Group
PO Box 6111d
Wyomissing, PA 19610-1118
(610) 927-0916

© Copyright 2009 ALERA Group, Inc. – All Rights Reserved

Copyright

© Copyright 2010 ALERA Publishing Group

All Rights Reserved. No part of this book may be reproduced or transmitted in any form or by any means electronic or mechanical, including photocopying, recording, or by any information storage or retrieval system, without written permission of the ALERA Publishing Group, Inc., except for inclusion of brief quotations in a review.

Published by ALERA Publishing Group, Inc. Wyomissing, PA 19610

Printed in the United States of America

ISBN 978-1-300-67630-0

About This Manual

Introduction

This manual is a training workbook for helping participants to understand the fundamentals of Overall Equipment Effectiveness (OEE). It is designed to provide clear and simple explanations of concepts, procedures and techniques that make up this program.

How To Use This Manual

There are four ways to find information in this manual:
1. The table of contents, which lists the sections of the manual
2. The table of illustrations, which lists the illustrations found in this manual
3. The map title at the top of every page which provides a quick reference as to which section the manual is open to
4. The footnote at the bottom of every page.

Description

The sections described in this manual include:
- Overall Equipment Effectiveness
- Manually Measuring OEE
- Operator OEEProgram
- Improving OEE

In This Manual

The sections described in this manual are located as indicated below:

Topic	See Page
Overall Equipment Effectiveness	1
Manually Measuring OEE	12
Improving OEE	19

© Copyright 2009 ALERA Group, Inc. – All Rights Reserved

Table of Contents

Table of Figures

Overall Equipment Effectiveness

Overview

Introduction	Overall Equipment Effectiveness (OEE) is a universal measurement that has been used worldwide for over 10 years. It is a formula to measure the efficiency of production line equipment. In short, OEE measures the ratio of first-pass acceptable product actually produced to the theoretical amount that could be produced under optimal conditions.
Why Overall Equipment Effectiveness	Overall equipment effectiveness is a measurement used to indicate how effectively machines are running.
In This Section	The topics described in this section are located as indicated below:

Topic	See Page
The Role of the Green Room Team in Using OEE	4
Losses Reduce Overall Equipment Effectiveness	4
Availability	8
Performance	10
Quality	11

Continued on next page

© Copyright 2009 ALERA Group, Inc. – All Rights Reserved

Overview, Continued

OEE vs. Efficiency	What do we mean by overall equipment effectiveness? Many people are familiar with the idea of "efficiency," which usually reflects the quantity of parts a machine or a person can produce in a certain time. OEE is different from efficiency in several ways.
Quantity Over Time Is Only Part of OEE	A machine's overall effectiveness includes more than the quantity of product it can produce in a shift. When we measure overall equipment effectiveness, we account for efficiency as one factor: • Performance In addition to performance, however, OEE includes two other factors: • Availability • Quality
Performance	A comparison of the actual output with what the machine should be producing in the same time.
Availability	A comparison of the potential operating time and the time in which the machine is actually making products.
Quality	A comparison of the number of products made and the number of products that meet the customer's specifications.
OEE Gives A Complete Picture	When you multiply performance, availability, and quality, you get the overall equipment effectiveness, which is expressed as a percentage. OEE gives a complete picture of the machine's "health"-not just how fast it can make parts, but how much the potential output was limited due to lost availability or poor performance. On subsequent pages, we will look more closely at these three elements and how they work together.

Continued on next page

 © Copyright 2009 ALERA Group, Inc. – All Rights Reserved

Overview, Continued

Effectiveness Focuses on the Equipment, Not the Person	Unlike some uses of the efficiency measure, OEE monitors the machine or process that adds the value, not the operator's productivity. When we measure OEE, we look at how well the equipment or process is working.
The Purpose of Measurement Is Improvement	Unlike some uses of the efficiency measure, OEE monitors the machine or process that adds the value, not the operator's productivity. When we measure OEE, we look at how well the equipment or process is working.
Improving Equipment Processes	Measuring OEE is not an approach for criticizing people. It is strictly about improving the equipment or process. Used as an impartial daily snapshot of equipment conditions, OEE promotes openness in information sharing and a no-blame approach in handling equipment-related issues. These key differences highlight the importance of OEE as a balanced measure that helps support improvement and profitability.

© Copyright 2009 ALERA Group, Inc. – All Rights Reserved

Overall Equipment Effectiveness

Losses Reduce Overall Equipment Effectiveness

Introduction

What makes machines less effective than they could be? The ideal, totally effective machine could run all the time (or whenever needed). It could maintain its maximum or standard speed all the time. It would never make defective products.

But most machines aren't ideal. They cannot run continuously. They cannot maintain maximum speed without problems...and they make defects.

Equipment-Related Losses

These problems are familiar forms of waste – they don't add value to the products. They reduce a machine's effectiveness, as measured by the OEE. The conditions that cause these machine problems are called equipment-related losses. Understanding the different types of equipment-related losses will give you a framework for applying OEE and participating in improvement activities to reduce the losses.

The equipment-related losses that are important for OEE are linked to the three basic elements measured in OEE: availability, performance, and quality.

Six Major Losses

There are six major losses that fall into the following three categories:
- Availability
- Performance:
- Quality:

Availability

Downtime losses include:
- Failures
- Setup time

Performance

Speed losses include:
- Minor stoppages
- Reduced operating speed

Quality

Defect losses include:
- Scrap and rework
- Startup loss

Continued on next page

Losses Reduce Overall Equipment Effectiveness, Continued

Basic Framework Although some companies link individual losses to different OEE categories, or add other losses that are especially significant for their operations, this basic framework is a useful starting point for many companies. Figure 1 below gives a visual image of the way in which these losses reduce the overall equipment effectiveness of a machine.

Figure 1: Impact of Losses on OEE

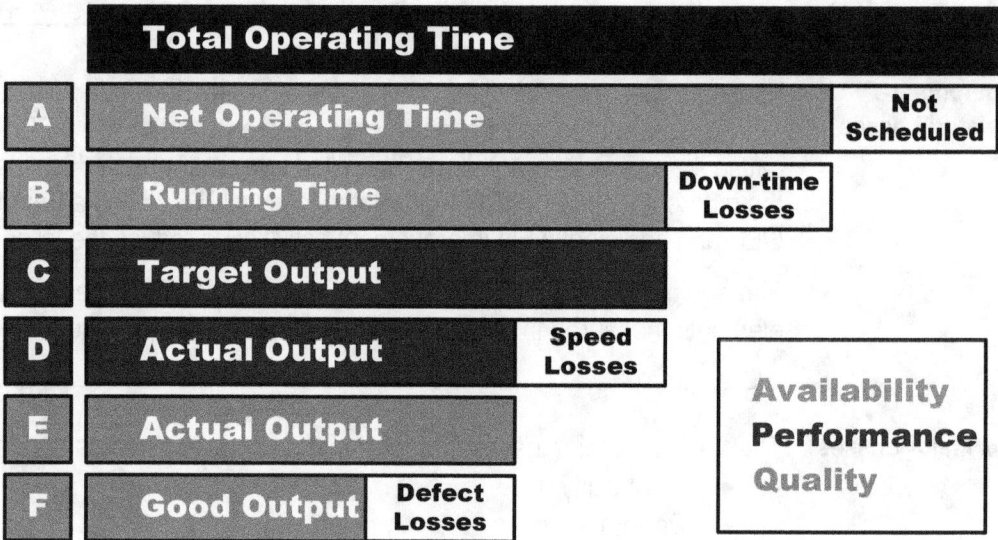

$$OEE = B/A \times D/C \times F/E \times 100$$

Visualizing OEE and the Losses Figure 1 makes it easy to see how OEE is derived from the three elements, expressed as fractions. Each pair of bars stands for one of the fractions--availability (B/A), performance (D/C), and quality (F/E). The fractions are often multiplied by 100 to turn them into percentages or rates.

Continued on next page

Losses Reduce Overall Equipment Effectiveness, Continued

Availability	Bars A and B represent availability. Unscheduled time shortens the total operating time,* leaving net operating time (A). But the machine is frequently down during some of that time, usually due to breakdowns and setup. Subtracting that downtime leaves the running time (B) in which the machine is making product.

Example

$$\frac{\text{Running Time}}{\text{Net Operating Time}} = \frac{300 \text{ minutes}}{400 \text{ minutes}} = .75 \text{ availability x } 100 = 75\%$$

Performance

Bars C and D represent performance. During the running time, the machine could produce a target output quantity (C) if it ran at its designed speed the whole time. But losses such as minor stoppages and reduced operating speed lower the actual output (D)

Example

$$\frac{\text{Actual Output}}{\text{Target Output}} = \frac{12,000 \text{ Units}}{20,000 \text{ Units}} = .60 \text{ Performance (x } 100 = 60\%)$$

Quality

Bars E and F represent quality. Of the actual output (E), most of the product is good output (F). But usually some output falls short of the specified quality and must be scrapped or reworked. Scrap is often produced during machine startup as well, lowering the yield from the materials.

Example

$$\frac{\text{Good Output}}{\text{Actual Output}} = \frac{11,760 \text{ Units}}{12,000 \text{ Units}} = .98 \text{ Quality (x } 100 = 98\%)$$

How Losses Impact Production

Figure 1 shows how losses to availability, performance, and quality compound to reduce the amount of good output a machine can produce during a shift. You can improve quality to raise the quantity of good output a little bit--but the total quantity won't rise dramatically unless you also improve both performance and availability.

Continued on next page

Losses Reduce Overall Equipment Effectiveness, Continued

OEE Formula	The formula at the bottom of Figure 1 shows how to multiply the three elements to get the OEE.

Example	.75 **X** .60 **X** .98 **X** 100 = 44% OEE {Availability) **(Performance)** **(Quality)**

© Copyright 2009 ALERA Group, Inc. – All Rights Reserved

Availability:

Downtime Losses

There are basically two types of downtime losses that we are concerned with:
- Failures
- Setup Time

Failures

Availability is reduced by equipment failures, which are a common occurrence in many plants. Machines used for production generally have lots of moving parts and various subsystems in which things can go wrong. When they do, the machine breaks down and stays down until repairs are completed.

Many of the causes of machine failure give warning signs before the machine actually breaks. In Chapter 4 we will look at how autonomous maintenance activities can help spot early trouble signs in time to prevent major breakdowns.

Setup Time

Availability is also reduced by the time it takes to set up the machine for a different product. In addition to changing the value adding parts, a changeover requires some preparation or make ready. It may involve cleaning and making adjustments to the machine to get stable quality in the next product. Too often, it also involves running around to find tools, parts, or people.

Continued on next page

Availability:, Continued

Other Losses to Availability

Failures and setup losses were the original losses counted as downtime that reduces availability. Some companies also track other losses as downtime, depending on what losses they are trying to improve. Cutting tool loss, startup loss, and time not scheduled for production are three other losses tracked as downtime at some plants.

Startup Loss

Startup loss is traditionally included as a defect loss, since its essence is the production of defective products during startup. However, startup loss involves lost time until good production can be stabilized, so it is logical to subtract it from available time as well.

Time Not Scheduled for Production

In some companies, when machines are stopped for meetings, preventive maintenance, or breaks, the time is considered "not scheduled" and is not counted in the availability rate (see Figure 2-5). Other companies recognize that even necessary activities like these reduce the available production time. They may decide to consider time "not scheduled" as a downtime loss that lowers the availability rate.

Performance

Speed Losses	Throughput loss due to lower than Control Point speed operation, including: • Reduced Operating Speed • Minor Stoppages

Reduced Operating Speed

Machines often run at speeds slower than they were designed to run. One reason for slower operation is unstable product quality' at the designed speed. In other cases, people don't realize that the equipment is designed to run faster. We will look in Chapter 3 at how to determine speed for the OEE calculation.

Minor Stoppages

Minor stoppages are events that interrupt the production flow without actually making the machine fail. They often occur on automated lines, for example when product components snag on the conveyor (see Figure 2-6). Minor stoppages can make it impossible to run automated equipment without someone to monitor it. These stoppages may seem like petty annoyances, but they add up to big losses at many plants.

Minor stoppages last only a few seconds, so we don't try to log the time lost. Instead, we include them in performance losses that reduce the product output.

Quality

Defect Losses

There are two types of loss associated with quality:
- Scrap and Rework
- Startup Loss

Scrap and Rework

Products that do not meet customer specifications are a familiar loss. Clearly, scrap that cannot be reused is a waste of materials. Even when products can be reworked, the effort spent to process them twice is a waste.

Startup Loss

Many machines take time to reach the right operating conditions at startup. In the meantime, they may turn out defective products while operators test for stable output. Some companies simply include this startup loss in scrap and rework; others single it out as a specific loss to track.

Quality problems happen when the optimum conditions do not exist at the moment when a person or machine works on the product.

Manually Measuring OEE

Overview

| Introduction | Measuring overall equipment effectiveness is an important way to monitor which losses are reducing the effectiveness of your machines. By tracking OEE on a regular basis, you can spot patterns and influences that cause problems for production equipment. Furthermore, measuring OEE allows you to see the results of your efforts to help the machines run better. This chapter offers guidance in measuring overall equipment effectiveness, including collecting and processing OEE data and reporting OEE results. |

| Closing the Feedback Loop | The process of measuring and applying OEE data should involve the people who use the machines. As operators, you are more familiar than other people with the equipment you operate, and you have a stake in helping it run well. Therefore it's logical for you to take part in collecting the data for calculating OEE

Just as important as being involved in data collection is receiving feedback on OEE results. An OEE chart cannot promote improvement if it doesn't get back to the shop floor. OEE is living information for improving equipment effectiveness. It should not be buried away in an office. |

| In This Section | The topics described in this section are located as indicated below: |

Collecting OEE Data

Introduction	It is important that Green Room Meeting participants know how to collect and calculate OEE data.
Defining What to Measure	Before you can begin applying OEE, you need to decide what machine and product data you will measure for the calculation. The basic items you will measure are the losses that reduce availability, performance, and quality. These will vary from plant to plant, but the Six Major Losses described pages 4 through 7 give a good framework to start from.
Downtime Losses	Downtime losses (lost availability) are measured in units of time (Figure 3-1). They include • failure and repair time • setup and adjustment time • other time losses that reduce availability
Failure and repair time	Failure and repair time includes all of the downtime until the machine makes the next good product. Some plants lump all breakdowns into one category; other plants may create several categories to distinguish between different types or causes of machine failures. The main thing is to standardize your approach so everyone can measure a failure event the same way.
Setup and adjustment time	Setup and adjustment time includes the time between the last good piece of product A and the first good piece of product B..
Other time losses	Other time losses include startup losses-similar to setup time losses--and any nonscheduled time the team chooses to subtract from the available time

Continued on next page

 © Copyright 2009 ALERA Group, Inc. – All Rights Reserved

Collecting OEE Data, Continued

Speed Losses	Speed losses (lost performance) are measured in units of product. You probably already track your output quantity. For OEE, you look at the difference between the actual output and the potential output if the machine consistently ran at the designed speed, or at the standard optimum speed for each product.
Minor Stoppages And Reduced Operating Speed	Speed losses include minor stoppages as well as reduced operating speed. Although minor stoppages are "events" like mini-breakdowns, they often occur so frequently that it is not practical to record the time lost during many frequent stoppages. For that reason, many companies monitor minor stoppages by tracking the output reduction they cause.
Actual Output Rate Compared To The Output Rate At The Designed Speed	To compare the actual output rate (machine speed) with the output rate at the designed speed, you have to know what the designed speed is. If this speed does not appear in the machine's documentation, you will need to set a standard, such as the fastest known speed at which the machine can run (this may vary for different products).
Figure 2: Scrap and Startup Losses Are Measured As Defective Output	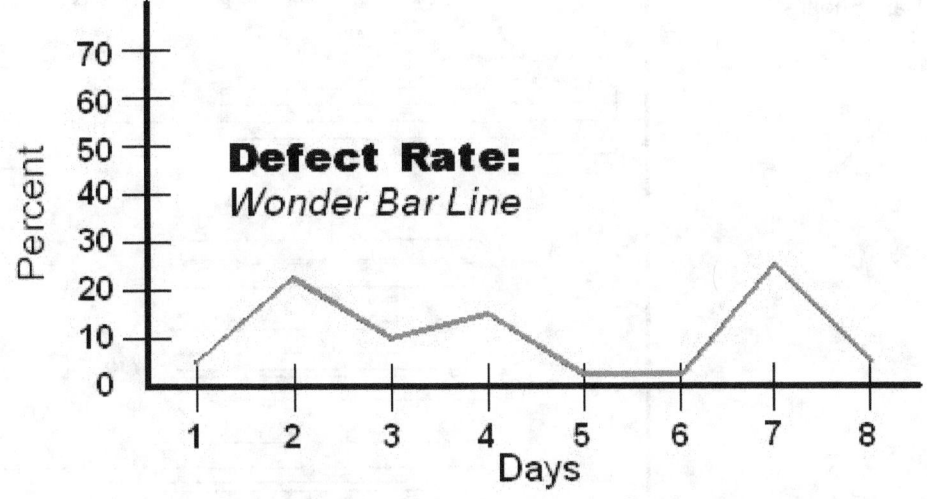

Continued on next page

Collecting OEE Data, Continued

Defect Losses

Defect losses (lost quality) are also measured in units of product output. This time, you are looking at the difference between the total actual output and the output that meets customer specifications (see Figure 2).

Defect losses include products that can be reworked as well as outright scrap. First-pass quality is the goal.

Making Data Collection Simple

The purpose of tracking OEE is not to make extra paperwork for operators. Most likely you are already collecting a lot of the data required for the OEE calculation. One well-designed form can make it easy to log the OEE data as well as other data you need to register during daily production.

Sample Form

Figure 3 and Figure 4 show a sample data collection form. Its creators used a simple approach for collecting data and making the calculation.. Side 2 is a trend chart showing OEE, availability, performance, and quality numbers

Figure 3: Sample OEE Data Sheet - Side A

Continued on next page

Collecting OEE Data, Continued

**Figure 4: Sample
OEE Data Sheet -
Side B**

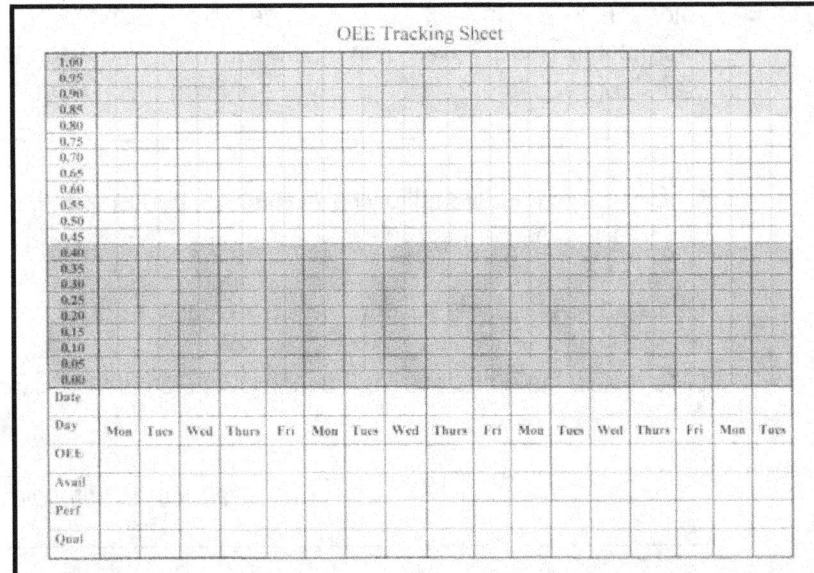

Making OEE Calculations

Processing OEE Data

After you collect data for OEE, you need to process the data to turn it into useful information. This involves doing the calculation, and also storing your data in a way that allows you to draw different types of information from it.

The OEE Calculation

OEE is calculated by multiplying availability, performance, and quality (multiplied by 100 to give a percentage rate).

OEE rate = Availability x Performance x Quality x 100

Let's review the equations for the individual elements of OEE.

$$Availability \quad = \quad \frac{Running\ time}{Net\ operating\ time}$$

The running time is the net operating time minus the downtime losses you decide to measure.

$$Performance \quad = \quad \frac{Actual\ output}{Target\ output}$$

For the OEE calculation, the target output is the quantity the machine would produce if it operated at its designed speed during the running time (see Figure 5).

$$Quality \quad = \quad \frac{Actual\ Output}{Target\ Output}$$

Continued on next page

Making OEE Calculations, Continued

Figure 5: OEE
Calculation And
Its Elements

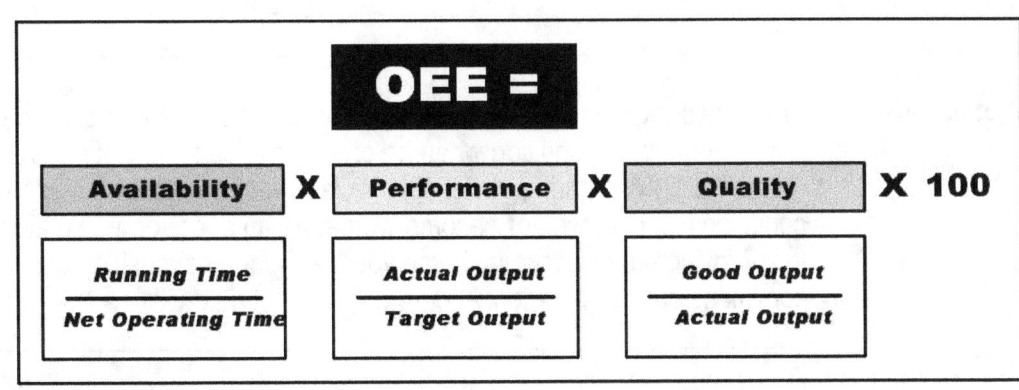

Improving OEE

Overview

Introduction	We measure OEE to monitor the condition of the equipment-similar to what a nurse learns about your condition when he or she takes a temperature or listens to a heartbeat. By comparing yesterday's or last week's result, we can see whether the condition has improved or become worse. As an operator, you play an important role in improving performance because you are in the best position to monitor machine conditions during operation.
OEE Measure Drives Improvement	This is the point "show us the opportunity for improvement. The OEE measure drives improvement.
Standardization	Standardization is the first step in improvement. OEE is a tool for standardizing the way you measure effectiveness. This standardized approach provides a baseline that helps you see where to focus improvement efforts.
Awareness Drives Improvement	Some improvement may happen just from the awareness that develops when you start measuring OEE. Sustained improvement, however, requires a dedicated approach, with management support. This chapter explores several approaches that can help improve OEE.
In This Section	The topics described in this section are located as indicated below:

5 – Why Analysis

Introduction

Have you ever had the experience in which someone fixed a machine problem, but the same problem happened again after a short time? In such cases, it often turns out that people have been treating the symptoms of the problem, but not dealing with its real, root cause. Until we address the root cause, the same problem will keep returning.

Determining Root Cause

5 – Why analysis is a useful tool that brings us closer to the root cause. As its name suggests, 5 – Why analysis involves repeatedly asking "why?" about the problem (it could be more or less than five times, depending on the situation). This leads us to look beyond the immediate effect--such as a broken drive belt--to see the factors that might be causing the effect--such as flaws on the pulley that make the belt wear out too soon.

Figure 6: 5 Why Analysis Technique

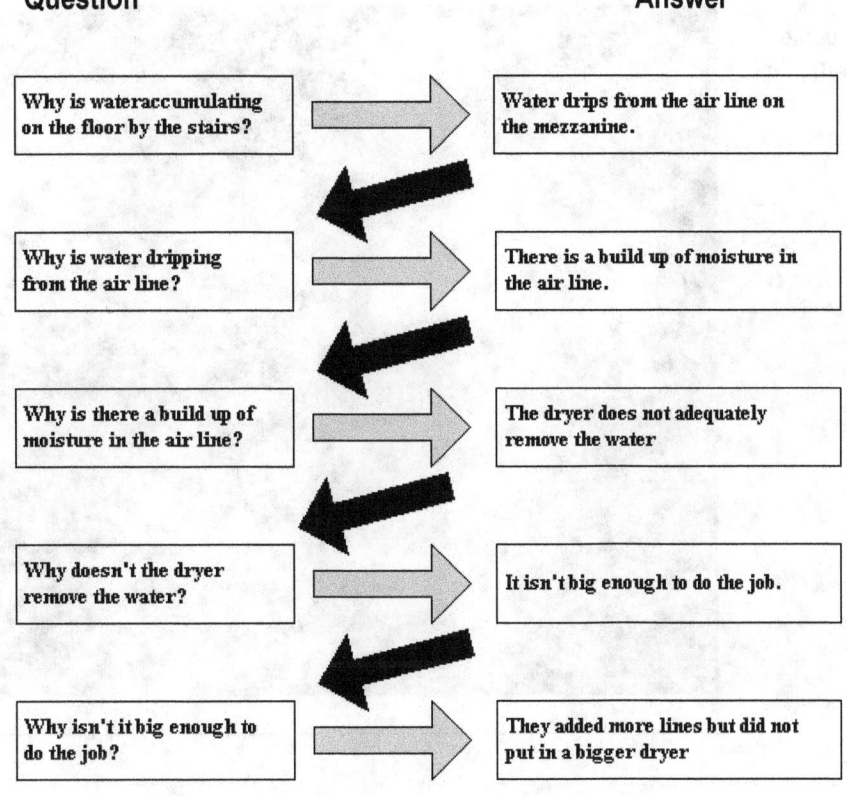

Question	Answer
Why is water accumulating on the floor by the stairs?	Water drips from the air line on the mezzanine.
Why is water dripping from the air line?	There is a build up of moisture in the air line.
Why is there a build up of moisture in the air line?	The dryer does not adequately remove the water
Why doesn't the dryer remove the water?	It isn't big enough to do the job.
Why isn't it big enough to do the job?	They added more lines but did not put in a bigger dryer

© Copyright 2009 ALERA Group, Inc. – All Rights Reserved

Basic Equipment Care

Introduction Basic Equipment Care refers to activities carried out by shop floor teams in cooperation with maintenance staff to help stabilize basic equipment conditions and spot problems early. Basic Equipment Care is one of the pillars of Total Productive Maintenance. It changes the old view that operators just run machines and maintenance people just fix them. Operators have valuable knowledge and skill that can help keep equipment from breaking down.

Fundamentals Of Basic Equipment Care In Basic Equipment Care, operators learn how to clean the equipment they use every day, and how to inspect for trouble signs as they clean (see Figure 7). They may also learn basic lubrication routines, or at least how to check for adequate lubrication. They learn simple methods to reduce contamination and keep the equipment cleaner. Ultimately, they learn more about the various operating systems of the equipment and may assist technicians with repairs.

Figure 7 Operator Involvement

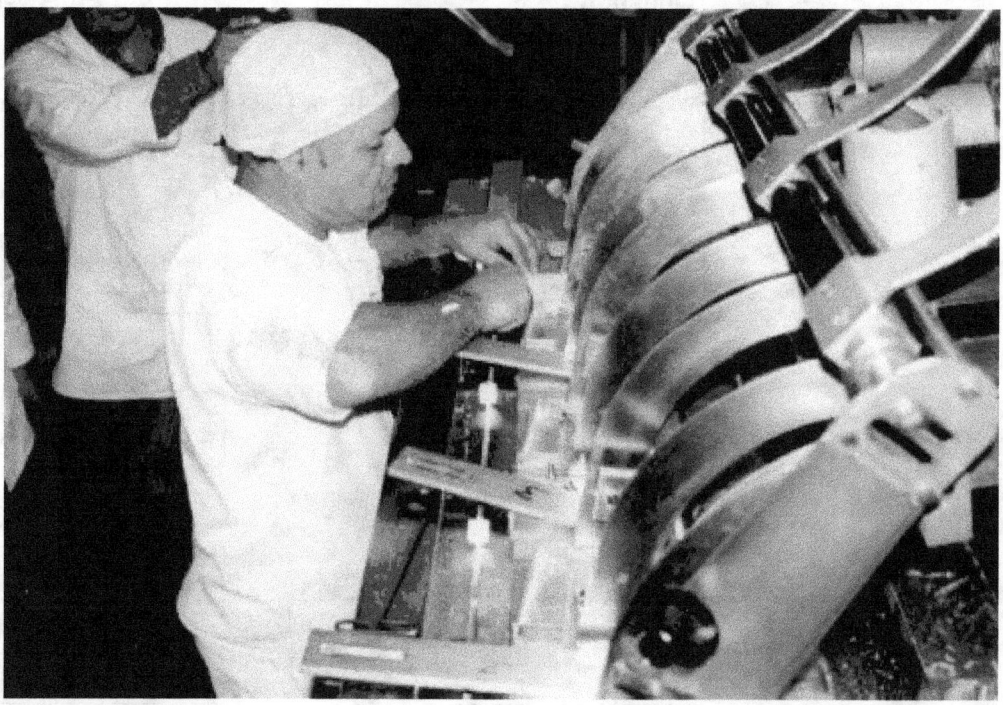

Continued on next page

Basic Equipment Care, Continued

Helping To Raise OEE

Basic Equipment Care activities are like exercise and regular health checkups for machines. Along with preventive maintenance, they help raise OEE by maintaining proper operating conditions, and stabilize it by detecting abnormalities before they turn into losses.

Seven Steps For Implementation

There are seven steps to implement a Basic Equipment Care program. These steps are as follows:

Step	Action
Step 1	Conduct initial cleaning and inspection
Step 2	Eliminate sources of contamination and inaccessible areas
Step 3	Develop and test provisional cleaning, inspection, and lubrication standards
Step 4	Conduct general inspection training and develop inspection procedures
Step 5	Conduct general inspections autonomously
Step 6	Apply standardization and visual management throughout the workplace
Step 7	Conduct ongoing Basic Equipment Care and advanced improvement activities

Focused Equipment & Process Improvement

Introduction	Focused equipment and process improvement is the Total Productive Maintenance pillar that deals most directly with improving equipment-related losses. If Basic Equipment Care and preventive maintenance activities are like exercise and health checkups, focused improvement is like an intense workout tailored to develop strength in specific muscle groups. Basic Equipment Care and planned maintenance improve OEE to a certain level, then help maintain basic operating conditions to stabilize OEE. To raise OEE beyond this stabilized level, companies apply focused improvement
Targeted Projects	In contrast to the ongoing activities of Basic Equipment Care and planned maintenance, focused improvement involves targeted projects to reduce specific losses. These projects are usually carried out by cross-functional teams that include people with various skills or resources an improvement plan might require. Depending on the target, a focused improvement team may include maintenance technicians, engineers, equipment designers, operators, supervisors, and managers
Eliminating Routine Problems	It's a good idea for companies to attain a basic "fitness" level with Basic Equipment Care and planned maintenance before launching focused improvement projects to address specific weaknesses. One reason is to eliminate routine problems (sporadic losses) so you have a clear view of difficult or more significant problems (chronic losses). Another reason is to avoid using a more expensive and time-consuming focused improvement approach for problems that could be addressed through less expensive Basic Equipment Care or planned maintenance.
Range of Approaches Needed	Focused improvement teams use a range of approaches to cut equipment-related losses. They may use 5 – Why Analysis as a starting point, but there are also approaches that address specific types of losses, such as setup losses and scrap. We will review approaches that deal directly with shortening changeover time and reducing losses from product defects. Finally, we will look at P–M analysis, an advanced version of root cause analysis that is used in focused improvement and quality maintenance.

Continued on next page

 © Copyright 2009 ALERA Group, Inc. – All Rights Reserved

Focused Equipment & Process Improvement, Continued

P-M Analysis

You may have experienced situations when you have to make repeated repairs and adjustments on a recurring problem. When a problem comes back, it is usually because the situation is not as simple as we originally thought it was. Our 5 – Why Analysis may have followed one factor to a deeper cause, but real life is complex and interrelated--several factors often work together to create a particular problem. P–M analysis is a tool for systematically uncovering and testing all the possible factors that could contribute to a chronic problem such as defects or failure.

The "P" in P–M analysis stands for "phenomenon"--the abnormal event we want to control. It also stands for "physical"…the perspective we take in viewing the phenomenon. "M" refers to "mechanism" and to the "4Ms"--a framework of causal factors to examine (Machine, Men/Women [operator actions], Material, and Method). P–M analysis is often spelled with a hyphen to distinguish it from abbreviations for preventive or planned maintenance

The essence of P–M Analysis is to look systematically at every detail so no physical phenomena, underlying condition, or causal factor is missed. Although product defects and equipment failures are the losses most often addressed, P–M analysis can be applied to any loss that involves an equipment abnormality

Continued on next page

Focused Equipment & Process Improvement, Continued

Conducting P-M Analysis

P–M analysis involves physically analyzing chronic losses according to the principles and natural laws that govern them. The basic steps of P–M analysis are

Step	Action
1	Physically analyzing chronic problems according to the machine's operating principles. This means understanding--in precise physical terms--what happens when a machine malfunctions. To do this, the team must first understand the physical standard for normal operation
2	Defining the essential or constituent conditions underlying the abnormal phenomena. This means understanding at the physical level what conditions exist when the machine doesn't work right. Examples include the position of the work or the temperature of a cutting tool.
3	Identifying all factors that contribute to the phenomena in terms of the 4M framework. This means examining the problem from several viewpoints to uncover factors the team might otherwise overlook.

Advanced Tool

P–M analysis is considered an advanced tool because this level of "detective work" requires more time, resources, and expertise than 5 – Why Analysis. For these reasons, focused improvement teams may save P–M analysis for complex or costly problems

Quick Changeover

Introduction	The focus of this program is on kaizen-type methods of improving changeover – small, inexpensive advances rather than big capital–expense innovations. Quick Changeover is based on the SMED system – *Single Minute Exchange of Die*.
The SMED System	The SMED system is a theory and set of techniques that make it possible to perform equipment set up and changeover operations in under 10 minutes. SMED was originally developed to improve die press and machine tool set ups, but its principles apply to changeovers in all types of processes.
The Benefits Of Quick Changeover For Your company	Quick Changeover changes the assumption that setups have to take a long time. When setups can be done quickly, they can be done as often as needed. This means Your company can make products in smaller lots, which has several advantages

No.	Advantage	Reason
1	*Flexibility*	Your company can meet customer needs without the expense of excess inventory.
2	*Quicker Delivery*	Small-lot production means less lead time and less customer waiting time.
3	*Better Quality*	Less inventory storage means fewer storage related defects. Quick Changeover also lowers defects by reducing set up errors and eliminating trial runs of new product.
4	*Higher Productivity*	Shorter changeovers reduce down time, which means a higher equipment productivity rate

Continued on next page

Quick Changeover, Continued

Benefits Of Quick Changeover For Workers

Quicker setups benefit you the worker by strengthening the company's competitiveness and thereby increasing job security. In addition, Quick Changeovers make daily production work go smoother because of:

- Simpler set ups result in safer changeovers, with less physical strain or risk of injury
- Less inventory means less clutter in the work place, which makes production easier and safer
- Setup tools are standardized and combined which means fewer tools to keep track of.

Introduction

There are three stages of Quick Changeover:

1. Separating internal and external setup
2. Converting internal setup to external setup
3. Streamlining all aspects of the setup operation

Separating Internal And External Setup

The most important step in implementing Quick Changeover is distinguishing between internal and external setup. By doing obvious things like preparation and transport while the machine is running, the time needed for internal set up, with the machine stopped, can be cut be as much as 30% to 50%.

Converting Internal Setup To External Setup

Further reducing times toward the single minute range (under 10 minutes) involves two important activities:

1. Reexamining operations to see whether any steps are wrongly assumed to be internal set up.
2. Finding ways to to convert these steps to external setup.

Note: Operations can often be converted to external setup by looking at their true function.

Streamlining All Aspects Of The Setup Operation

To further reduce setup time, the basic elements of each setup are analyzed in detail. Specific principles are applied to shorten the time needed, especially for steps that must be done as internal setup, with the machine stopped.

Appendix

OEE – Overall Equipment Effectiveness

Level		Availability Downtime Losses (Waste from stoppages)	Performance Speed Losses (Waste from poor performance)	Quality Rate Defect Losses (Waste due to defects)
1	Loss Reduction (**Note**: Measure these:)	• Breakdown & Stoppage Losses • Set Up & Adjustment Losses • Forced Idling Losses • Planned Downtime Losses	• Minor Stoppage & Idling Losses • Slowdown Losses • Reduced Capacity	• Scrap & Rework Losses (B&R) • Minor Adjustment Losses • Test Run & Adjustment Losses • Start Up Losses • Yield Losses
2	Focused Improvement (Programs and methodologies)	• Quick Changeover • MTBF Analysis • MTTR Analysis	• Analysis of minor stoppages • Phenomenon-Mechanism (P-M)Analysis • Equipment Performance Analysis	• Standard Operation Monitoring • Precision Management • Mistake-proofing • Quality Component Management
3	Increased Capacity (OPI) (These can be accomplished after you know what is happening and you have supporting programs in place)	• Reduced Number Of Employees • Automated Processes • Design For Reliable Operation	• **Expanded Equipment Capacity** • Compressed Processes • Increased Speed	• Increased Process Capacity • Optimum Conditions Set • Concurrent Engineering

© Copyright 2009 ALERA Group, Inc. – All Rights Reserved

ALERA Books and Programs

Topic	See Page
Books	30
Training Programs	34
Services	35

ALERA Group Website

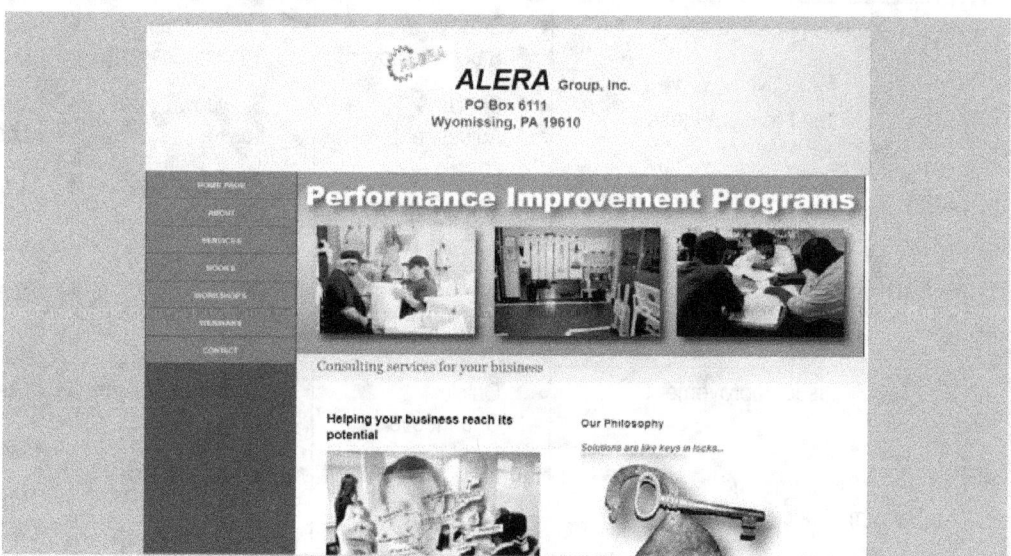

http://www.aleragroup.com

 © Copyright 2009 ALERA Group, Inc. – All Rights Reserved

Books From ALERA Publishing Group

5S Topics

Planning & Implementing 5S

Paperback	$21.52
Hard Cover	**$36.53**

The Planning & Implementing 5S program shows you how to organize a Performance Improvement Steering Team, how to analyze the workplace, how to plan a facility-wide improvement program, and how to sustain your efforts.

5S Workbook

Paperback	**$14.48**

The 5S workbook is the companion to Planning and Implementing 5S by Brice Alvord. It provides the tools used in the ALERA workshop.

Auditing Your 5S Program

Paperback	**$12.24**

The 5S audit is critical to the success of your 5S program. It is often overlooked or considered an unnecessary extra expense. The audit validates the accountability of the target area owners for complying with 5S plans. Without the audit, the program slowly withers away and becomes ineffective. A close look at 5S failures will reveal a lack of or an ineffective auditing program. This book explains how to conduct a proper 5S audit.

Strategic Planning For 5S

Paperback	**$30.51**

Strategic Planning for 5S is intended for managers and steering committee members who are considering the implementation of a 5S program. It shows how to apply Strategic Thinking to the 5S planning and implementation process and develop a strong business case for change

Continued on next page

Books From ALERA Publishing Group, Continued

Training

How to Train People On The Job

Paperback $18.53

NEW and REVISED workbook for Training On the Job Trainers. Covers adult learning theory, why shadow training does not work, how to perform a simple job/task analysis, how to develop trainer's guides and teach using the Four Step Method

Performance Based Training

Paperback $21.95

This book is intended as a guide to Performance Based Instructional Design. It covers how to conduct an effective Training Analysis including Job/Task Analysis, how to identify and define realistic competencies and instructional objectives and how to organize analysis data into a performance based training design. The book also explains how to develop important training documents including trainer's guides and lesson plans, participant manuals, and support materials including training and job aids and other media. Performance Based Training: Building a Competent Workforce is intended for the training professional as well as those people who have been given a training assignment. It is also a good reference for managers and supervisors to help them build a stronger workforce and to support company training efforts.

Advanced Instructional Design

Paperback $21.95

Advanced Instructional design focuses on the steps required to develop a performance based training design. Chapters include information conducting a Job Task Analysis and the Design of the training program. Other topics include defining competencies, conducting a DACUM, writing performance based objectives, developing criterion tests, Sequencing training elements, and writing a training blueprint. This book does not cover the development of training materials that will be addressed in another book yet to be published.

Continued on next page

 © Copyright 2009 ALERA Group, Inc. – All Rights Reserved

Books From ALERA Publishing Group, Continued

Training, Continued

One Point Lessons

Paperback $11.94

This book is a training workbook for developing One-Point Lessons. It is designed to provide clear and simple explanations of procedures and techniques to quickly create short, cost effective training materials.

Other Topics

Operations Performance Improvement

Paperback $21.59

Hard Cover $36.30

Fundamentals of Operations Performance Analysis shows your OPI team how to develop effective solutions to persistent performance problems. Your team will learn how to isolate and understand the root cause of defects and failures within equipment mechanisms and peripheral systems. They will learn how to apply a systematic approach for effectively controlling those causes.

Strategic Thinking

Paperback $13.16

Strategic Thinking for Operations & Projects focuses on how to build a strategic based business case for change. It is a powerful communications tool for getting projects approved.

Continued on next page

Books From ALERA Publishing Group, Continued

Other Topics, Continued

Fundamentals of Project Management

Paperback $16.38

Project Management Fundamentals covers the fundamental skills required to plan and implement project. It is intended for new project managers and managers with little or no project management experience.

Reliability Team

Paperback $27.08

Improving Operations Performance with a RELIABILITY TEAM explains what a Reliability Team needs to know in order to function properly. This book covers how to form a team, write a charter, how to run the team once it is created. this book provides a foundation for teamwork and continuous improvement activities.

Training Programs From ALERA Group

How To Train People On The Job

A 2 Day Hands-on Workshop that teaches your participants how to conduct On Job Training using the Four Step Method of Instruction.

Planning & Implementing 5S Workshop

A 3 Day Hands-On Workshop that teaches your participants how to plan and implement a basic 5S program. They will actually begin implementing 5S in a target area of your facility.

Team Based Problem Solving

A 2 Day Hands-on Workshop to teach your teams how to work together to identify and solve real problems in the workplace. Teams will address n actual problem and apply the tools to solve it.

Project Management Workshop

A 2 Day Hands-on Workshop that teaches the fundamentals of project management. Participants develop all of the elements of a project.

Overall Equipment Effectiveness

A 1 Day Hands-on Workshop to teach your participants what Overall Equipment Effectiveness is and how to calculate it accurately.

Try Z Seminar

A 2 ½ to 4 Day Hands-on Workshop from QCDSM Systems, Inc. See QCDSM Program Information on page **Error! Bookmark not defined.**.

Services From ALERA Group

Introduction	The ALERA Consulting Group exists to assist you in improving all areas of performance in your organization. We have a variety of state of the art tools and processes to help you identify performance needs and relate them to business practices and strategies.
Strategic Thinking	ALERA helps you develop strategic thinking in your organization; We conduct a strategic thinking workshop for selected members of your management team. We coach them through the application of the Strategic Thinking model to help them develop comprehensive and effective business cases for change within your organization.
Team Building	ALERA helps you design and deliver the right customized team development program, team building event, corporate retreat, or executive retreat that will improve your team's effectiveness, collaboration skills, and team-based results.
New Leader Assimilation	ALERA's Leader Assimilation program is based on the process designed by Kaiser Aluminum. Kaiser discovered that it normally took an incoming manager six months to become fully productive. The process was designed to reduce this amount of organizational down-time.
High Impact Change Management	ALERA's High Impact Change Management is a 3 phase organizational design program that assesses the organization with a performance audit, rationalizes change, and develops a comprehensive design. ALERA provides: design team formation and training, strategy focused design, alignment of the organization for maximum effectiveness, and building an empowered organization.
Asset Effectiveness (Focused Equipment Improvement)	ALERA helps your Operations Improvement Team develop the skills to address chronic equipment problems that hinder your profitability and overall performance. We provide workshops o help your team succeed. We evaluate your team's performance and coach individual members and your management team.

Continued on next page

Rev. 2 06/23/09   © Copyright 2009 ALERA Group, Inc. – All Rights Reserved   35

Services From ALERA Group, Continued

Training Analysis ALERA conducts or teaches your team to conduct a variety of training analysis including: training needs analysis, job/task analysis, cost benefit analysis, Our training professionals conduct training effectiveness audits, subject matter interviews, and individual performance evaluations.

Workplace Organization (5S planning and implementation) ALERA helps you analyze your needs, design a program, plan 5S implementation, evaluate the progress of your program, perform 5S audits, coach your management team on implementation problems and opportunities.

Project Management ALERA has experienced project managers who can assist you with keeping your performance improvement project or training project on schedule and under budget. We address your organizational needs and support consulting efforts with comprehensive training programs for your team as needed.

Technical Writing and Instructional Design and Development ALERA can provide you with technical writers to help you develop standard operating procedures, lockout/tagout procedures technical documentation, training manuals, detailed process sheets.

NOTES

Overall Equipment Effectiveness

NOTES

NOTES

www.ingramcontent.com/pod-product-compliance
Lightning Source LLC
Chambersburg PA
CBHW081244180526
45171CB00005B/534